A (How Not To) Crash Course in Systems Engineering

A Special Acknowledgment

Thank you to my many mentors who have guided me in becoming a successful engineer. The content of this book was made possible by you putting me on the right path.

Thank you to my father, who never asked for thanks while giving me the tools I needed to nurture my curiosity.

To the ambitious and curious minds of the future. This book is dedicated to you. You inspire me.

“Somewhere, something incredible is waiting to be known.”

-Carl Sagan

Table of Contents

Foreword

As an early-career or aspiring engineer, it's quite possible you're thrilled to have the opportunity to make a difference in the industry! Perhaps you want career growth through technical progression. Perhaps you want to be a versatile engineer who can work in a role different from the one you initially specialized in. Perhaps you just want to understand how to manage a product through a rigorous lifecycle. The important takeaway is that there is always more to learn and improve on. This is the basis of innovation.

When I first started my career, the concept of **systems engineering** and **systems thinking** was completely foreign to me. As I progressed, it became increasingly apparent just how much the *puzzle* of designing and innovating complex things came together once I applied this philosophy to my work.

I had always had a fascination for difficult things, things that flew, and things that operated effectively. What continued to impress me was the fact that HUMANS did all of those things. Humans built engines from natural phenomena. Humans discovered flight and built aircraft capable of traveling great distances that were once seemingly impossible. Humans put humans on the moon. Humans manipulated their surroundings and transformed the world's raw materials into technological marvels.

The act of creation is one of humanity's most extraordinary gifts. We can envision and bring to life ideas in vivid detail. Our imagination allows us to transform abstract "thought-forms" into tangible reality. This is a fundamental aspect of who we are as humans. Consider the work of an engineer, for instance. When they embark on a design project, they are essentially crafting a narrative. They're crafting a story filled with challenges, risks, and opportunities. In this narrative, there exists an end-user, someone grappling with a specific problem that demands a solution.

The engineer translates this storyline into actionable plans and turns those plans into an innovation.

It's quite apparent by the ingenuity that brought us to this point that humans are addicted to innovation. Simply put, we make better stuff, and we don't quit. Now, what's especially fascinating is that progress has always been a team sport. The question that always stood out to me was HOW did we progress like this? We can barely get along! It's hard to get people to work together as a team, and it's especially difficult to get people to collaborate and share complicated ideas. So, how do we get folks to do that and build complex engineering marvels?

The hard truth is that we're learning how to do this every day, but we've just happened to get pretty good at it over hundreds of years of trying (often through catastrophic failures). During the 1940s and 1950s, the term **systems engineering** was coined by Mervin J Kelly of Bell Labs. The elusive concept of **systems engineering** was the first step to condensing the chaos of innovation.

Eventually, I, now a principal engineer, have come to realize the significance of systems engineering and systems thinking in modern engineering and why they're so important for both delivering a robust product and advancing in an engineering career. It teaches you how to lead, negotiate, and most importantly, integrate your way to the best solutions. It's truly why things in society these days are significantly more reliable.

Now that I've teed us up, I'm going to break it down for you in as simple a way as possible. First things first, this is not a textbook. So what is the purpose of this book? A lot can be said about systems engineering. I, in particular, care about the core essentials that help create productive, highly effective engineers who can hit the ground running. Being able to streamline yourself into a role is one of the best qualities an engineering candidate can have. With basic systems engineering knowledge, I personally believe anyone can become an engineer (and a highly effective one, at that). These are basics that matter! And at the end, I'll show you how you can take all these things to create a winning resume and perform well in your interviews. That is why I wanted to write this book.

This is a "How Not To Crash" course. Let's get started.

Chapter 1: What Is Systems Engineering?

Intro

Imagine a busy restaurant where the chef is busy crafting gourmet dishes, but no one communicates with the servers. The servers, in turn, don't coordinate with the host, leading to double-booked tables or customers waiting far too long to be seated. In the back, the dishwasher calls out sick, and no one notices until dishes start piling up, bringing the kitchen to a halt.

The ingredients may be excellent, and the staff may be skilled, but without proper coordination, the entire operation falls apart. Orders get delayed, tables sit empty, and customers leave angry, writing horrible Yelp reviews.

This is what happens in complex engineering projects without systems engineering. Even when individual teams excel at their roles, developing state-of-the-art designs, performing analyses, or manufacturing components, things can go awry if there’s no overarching strategy or continual assessment of risks. Misaligned goals, overlooked details, and unanticipated bottlenecks can derail the entire project.

Systems engineering addresses this challenge by focusing on the relationships and dependencies within the system. It ensures that all the moving parts are synchronized and aligned with the ultimate bigger picture, avoiding the kind of disarray that leaves your “restaurant” in shambles. How do we fix something complex like this? Well, a systems engineer's mindset is that of an integrator. With that in mind, we integrate.

Before service begins, a systems engineer would work with each **stakeholder** to map out processes and identify potential bottlenecks. The host, servers, kitchen staff, and dishwashing team would all have clear workflows, and mitigation plans would be in place for the absence of a dishwasher, sous-chef, or server. Communication channels would be established to address issues in real time, ensuring the entire operation runs smoothly. To ensure that the service is ready, the systems engineer would develop requirements for the team to meet. Upon verifying these requirements, the restaurant is ready to serve! This could be seen as overkill, but it’s the simplest, tried-and-true way to engineer a system's complexity into something far more ideal.

Another example:

The Boeing 787 Dreamliner is one of the most reliable and leading long-haul aircraft. It’s a (relatively) lightweight aircraft that’s more fuel-efficient than its predecessors; however, its development journey was NOT smooth.

Boeing made the ambitious (penny-pinching) decision to outsource much of the Dreamliner's production to suppliers around the globe. The idea was to speed up development and reduce costs, but the complexity of managing such a dispersed supply chain quickly spiraled out of control.

For example, one supplier delivered incomplete fuselage sections that didn't align correctly with other components, requiring extensive rework. Another issue arose when avionics subsystems from different vendors couldn't communicate properly, leading to delays in testing and integration. These problems stemmed from a lack of oversight in systems engineering.

Eventually, Boeing tightened its systems engineering practices for the Dreamliner, emphasizing better communication, rigorous validation, and closer coordination between teams. Once these processes were in place, the Dreamliner turned into one of the most successful aircraft in aviation history and one of my personal favorites. However, the Boeing 737 Max situation shows us, once more, what happens when proper systems engineering is neglected.

The Max was rushed to market with inadequate oversight of systems engineering, resulting in a poorly integrated software system. This was the now-infamous MCAS (Maneuvering Characteristics Augmentation

System) that pilots weren't properly informed about, which was approved with a glaring single-point-failure mode. To be a little more specific, there was a design flaw with an "Angle-of-Attack" sensor that anomalously interpreted normal takeoff conditions as the aircraft stalling. This would trigger MCAS inadvertently and force the aircraft's attitude down. Pilots were not trained on how to override the MCAS in a scenario like this.

The lack of thorough validation and cross-disciplinary coordination led to catastrophic failures, resulting in multiple crashes, many deaths, and the grounding of an entire fleet of planes. By prioritizing shareholders and cost-cutting over proper engineering integration, Boeing went from a household name in aviation to being an embarrassment to the industry. This ultimately weakened trust in commercial aviation as a whole (for example, your uncle who likely says he would rather drive than take one of *those* flights).

No matter how advanced or familiar a design might seem, proper integration depends on robust systems engineering practices that ensure safety, reliability, and cohesion throughout the project.

The Systems Engineering V

One hallmark of systems engineering and systems thinking is its focus on the entire lifecycle of a project. An engineering project is a temporary development with a defined start and finish.

A project's lifecycle is often depicted using the **Lifecycle V** model. At the beginning of the process, systems engineers work with stakeholders to define the project's goals and constraints. This process is performed to understand the system's purpose, the environment in which it will operate, and the challenges it must overcome.

Let's make this clear: *any* problem or project can be modeled as a complex system consisting of subsystems, interacting elements and relationships, and constraints.

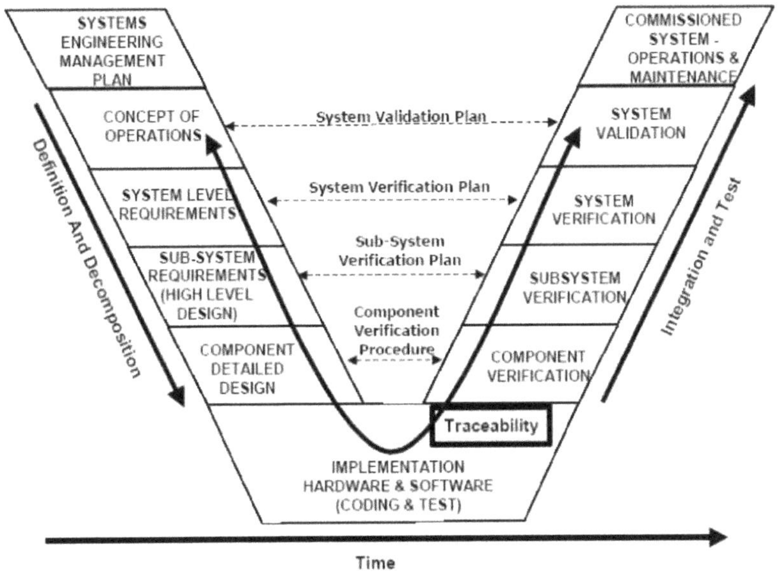

Once the goals are clear, the project moves into the design phase. Here, the high-level requirements are broken down into detailed specifications for each subsystem. For an airplane, this might mean defining the thrust requirements for the engines, the weight limits for the fuselage, and the responsiveness of the flight control system. Systems engineers also define the interfaces between subsystems to ensure that they can work together seamlessly.

Implementation begins with the components being built and tested individually. Once the parts are assembled into a complete system, the real work of integration begins. Each interface is verified, previously held assumptions are rigorously challenged, and potential failure modes are assessed as part of this rigor.

Finally, the system enters the verification and validation phase. Verification answers the question, “Did we build the system right?” Validation asks, “Did we build the right system?” These steps are critical for ensuring that the final product not only meets its specifications but also fulfills its intended purpose.

At its heart, systems engineering is about perspective. It requires you to zoom out and see the forest (the entire system) even as you dive into the details of individual trees with your team. For example, increasing the power of an airplane's engines might improve its speed, but it could also increase weight, reduce fuel efficiency, and strain other subsystems. Systems engineers weigh these trade-offs to ensure that each decision aligns with the project's big picture goals.

A point that I would like to really drive home is that EVERY engineer in a development or sustainment program is a systems engineer. If you're operating in a system, you are a systems engineer in some capacity. Understanding how to utilize the systems engineering "toolkit" is a major key to success in your engineering career.

Design Reviews

Design reviews are incremental opportunities to bring together all project stakeholders and assess the design's maturity throughout the lifecycle. These reviews are typically large, interdisciplinary presentations (often with free lunch) that mark milestones. The flow typically looks like this.

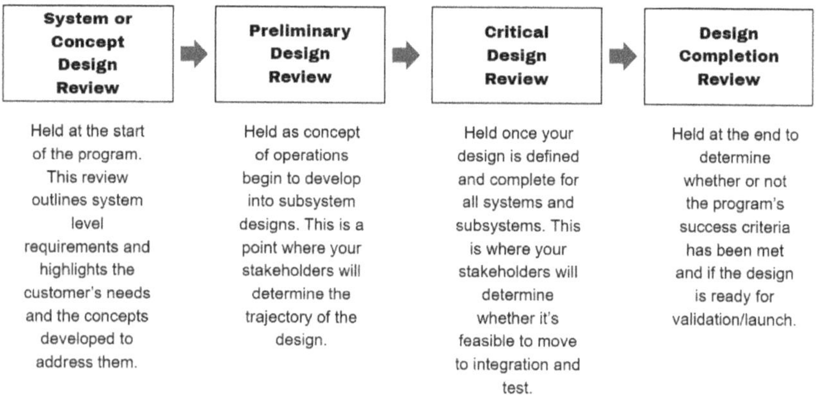

It should be stated that every organization operates differently. Some of these review names may look different, and sometimes you can expect an organization to sprinkle in additional reviews based on their development rhythm.

Soft Skills

I also briefly want to touch on soft skills. I won't dive into too much detail here, as there's quite a bit to say about each, and I would like to focus on why they're relevant to systems engineering. There are a few key characteristics that systems thinkers tend to have that make them more effective at integration, and as we progress through this book, you will understand why. These characteristics are:

- **Communication**. Responsible engineers are experts in their domain, which means they're not experts in another domain. You can't expect your brilliant fluid engineer to be as proficient with electronics engineering. In the same way, you can't expect your electronics engineer to be a fluid engineering expert. How do we get them to collaborate effectively to solve problems? Simple, you communicate. It's super easy to silo yourself as an engineer, so to avoid that pitfall, you integrate.

Communication is one of those skills that you can practice just through exposure. Obviously, everyone has different communication styles and approaches to bringing people together. The fact of the matter is, to be an effective engineer, you have to be willing to play nice with other engineers. There's an element of social charisma that can really assist here. There's also an element of understanding that sometimes, you will need to be annoying to get your team's attention on emergent items.

If you aren't a very good communicator, what I've found is that watching movies and TV shows is remarkably helpful for this. I, personally, had to work at becoming a good communicator. Actors are best at their craft at evoking emotion through communication and gesture, so this is my personal winning formula: watch movies or shows with a leading actor you like and observe how they carry themselves in interactions with other characters. Practice some of their communication mannerisms or gestures. You may find that your ability to evoke emotion will improve as well.

- **Time management.** As a systems engineer, you will have to prioritize your time across projects. Some integration items will be hotter than others. An unfortunate side effect of being a systems engineer is being inundated with various "contexts" to think about. Sometimes it's very important to recognize that you're overburdened and to prioritize your time accordingly. This will make you a much more efficient integrator.

- **Tracking interfaces.** When a product moves through its lifecycle, you will consistently notice shifts and changes in the interfaces between design groups. For instance, if a software update were made, there's a pretty good chance it would affect the electrical team downstream. If there were an adjustment made to how an airplane's fuselage is developed, well, that may drive a change to the way aerodynamic margins were initially assessed. We'll dive deeper into the significance of this skill, in particular, in the next few chapters.
- **Taking ownership.** While this may be a controversial view, arrogance can have a place in engineering. There's an art to it, however. You have to learn to be arrogant about the things YOUR system needs to progress. Your ultimate objective as an integrator is to form educated opinions in order to own large decisions. Being an "owner" of a subsystem or a system means you assert yourself as the bearer of the specification and the design. This will establish you as the steward of the system's lifecycle. During an integration cycle, the systems engineer becomes the technical owner of the integrated product. For instance, I was the systems engineer for an avionics group. This meant that I was the technical owner of the integrated avionics system.

Beyond just making you a better engineer, being a technical owner is also a great way to become *the person* that people can come to for a particular need. It makes things much more efficient if a stakeholder doesn't have to go fishing for information in a nebulous aether. They can simply walk over to your desk and talk to you about your system as needed. (Also, it stands to be said that this is a great way to gain visibility in your career. You can brag about this.)

A piece of taking ownership stake is also about understanding what you DON'T know. Ask questions. Ask lots and lots of questions. You don't have to be an expert on everything.

Thankfully, these skills can and will be honed on-the-fly. If you are a budding or aspiring engineer and you feel as though you have a lot to learn ahead of you, that is OKAY! In fact, a large part of engineering is understanding that you always have plenty to learn and refine. Never stop being a learner.

In the next chapter, we're going to dive deeper into the lifecycle of systems engineering with examples.

The How-Not-To-Crash Takeaways from Chapter 1

- Systems thinking is about the big-picture perspective and the intertwining of many stakeholders.
- Systems thinking involves considering a product's lifecycle and maturity.
- Every engineer is a systems engineer in some capacity.
- You're going to be seeing the word "integration" a lot.
- Always be open to learning.

Chapter 2: The Lifecycle of Systems Engineering

Designing a System, One Step at a Time

We'll start with a simple design.

Imagine someone asking you to prepare a sandwich for them. Every system begins with a problem or a necessity. For a sandwich, that necessity is to satisfy a customer's hunger or cravings.

It seems straightforward initially, but as soon as you factor in preferences, like how much peanut butter, the type of jelly, or whether the bread is toasted, it quickly becomes clear that vague instructions are not enough (unless the customer just doesn't care what kind of sandwich they get).

They might say, "Make me a PB&J sandwich," but the result could still vary wildly depending on who's making it. What if they don't like strawberry jelly and would prefer grape jelly instead? Well, to get the sandwich *exactly* the way the end client wants it, there needs to be clear, measurable requirements:

Vague: "Make me a sandwich."

Precise:

- **Requirement 1**: "The sandwich shall have two slices of whole-wheat bread."
- **Requirement 2**: "The sandwich shall have one bread slice spread with 0.8 oz of peanut butter."
- **Requirement 3**: "The sandwich shall have one bread slice spread with 0.9 oz of grape jelly."
- **Requirement 4**: "The sandwich shall be assembled with the peanut butter and grape jelly facing inward."

Now obviously, you don't have to necessarily talk like this in a casual setting, but the takeaway is that the keyword **"*shall*"** turns a vague idea into an enforceable standard. In the systems engineering methodology, this level of precision ensures everyone involved, from designers to testers, knows exactly how to verify that the criteria have been met for delivery. Let's continue to design our sandwich like we would a complex system.

I want to make it clear that developing a system requires many requirements across many different subsystems. When you develop high-level requirements, you can always break them down to the component level. The best way to do that is flowdowns.

If you have a "system-level" requirement, you can build a structured set of lower-level requirements that collectively ensure that the higher-level requirement can be satisfied. Those are the lower-level requirements that then become the basis for design, procurement, and verification at the component or *subsystem* level.

Let's use the example of Requirement 3. "The sandwich shall have one slice of bread spread with 0.9 oz of grape jelly."

To properly develop or source the jelly for this sandwich, you would need to write requirements for that subsystem. We'll call this the Grape Jelly Subsystem. That requirement specification could include requirements for jelly composition, viscosity, sugar content, packaging, storage temperature, shelf life, and mass measurement tolerances. Each of these lower-level requirements would flow up to support the 0.9 oz application requirement at the sandwich (system) level. The best approach is to collaborate with subsystem subject-matter-experts (SMEs) to write this specification. As a systems engineer, you have to understand where to rely on the responsible SMEs. Below is a figure depicting how this system would flow in our context.

Challenge: Can you identify more subsystems or components that we could develop from this?

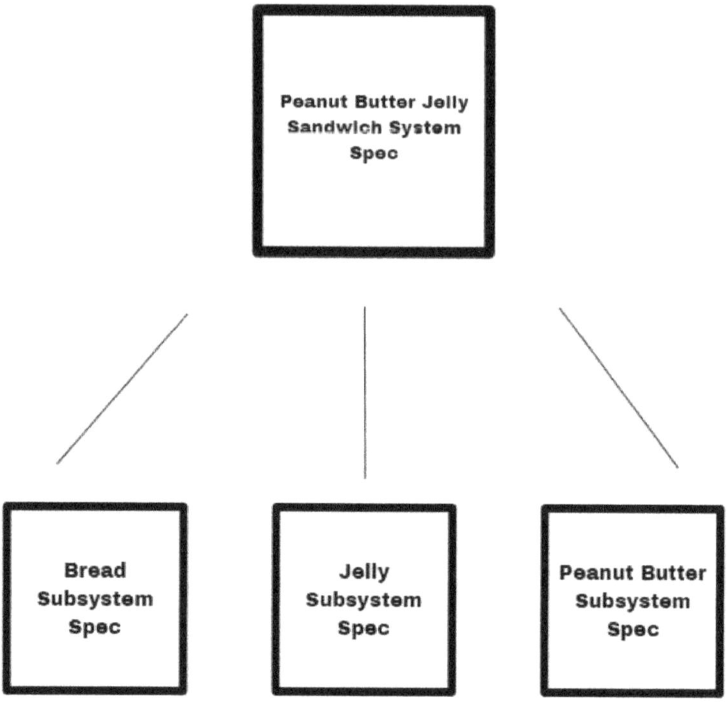

If you are vertically integrated, you would use those subsystem requirements to design and develop the subsystem in-house; however, that's not how many firms operate. A lot of the time, you can't develop everything in-house. In a case like this, you would work with a supplier. You and the SMEs would coordinate and enter into a contract with a supplier to design a product that meets these requirements. This helps with traceability inside the design. If you change one requirement, you may be affecting another. Mindful traceability helps maintain the interconnected nature of systems and subsystems.

On Traceability and Impact Analysis

Impact analysis is a critical function, particularly when introducing design changes or iterations. Even a seemingly minor modification can have broader system implications. For example, replacing standard fasteners on a component with grounding screws may seem straightforward, but downstream effects can be significant. This change could require new hole patterns and compromise structural integrity, necessitating an engineer to conduct a requirements scrub.

Without solid traceability in your system, this relatively simple change could balloon your labor hours and your project's cost. Nobody wants a situation where you're spending 40 hours on a very manual impact analysis. Often, PLM tools will have built-in features for this exact case. When you break your requirements down, you establish parent/child relationships among them as they flow through subsystems and components.

Now, for the sake of this chapter, we will focus specifically on our designated system-level requirements for the preliminary design. Let's look back at the Lifecycle V. You can continue to break the design down into more and more requirements to fine-tune your sandwich. The point at which you reach the bottom of your Lifecycle V, you're ready to build a prototype. This is the Integration and Test stage, where you'll share your specification and design with your technicians and can start verifying and validating.

Typically, around this point, there will be a critical design review to bring in all stakeholders and get their input.

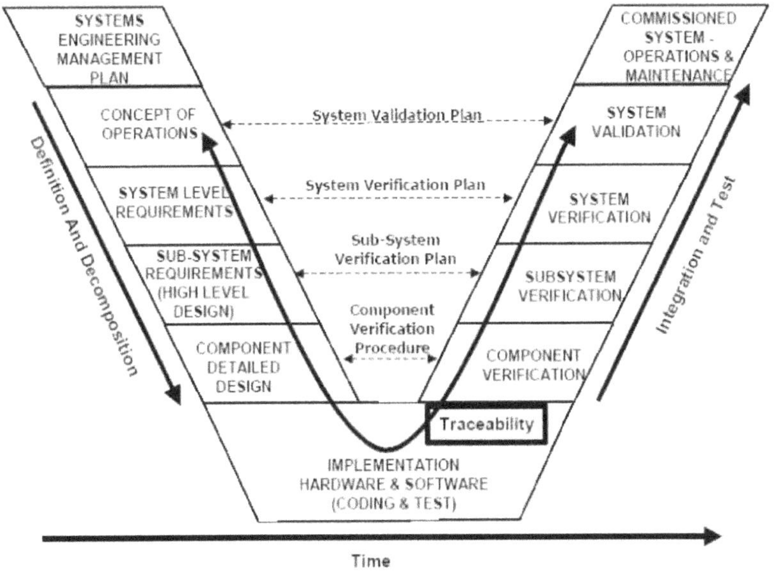

Verifying and Validating a System, One Step at a Time

At this stage, your next step is to determine how to verify the requirements you've developed.

To verify these, we must first determine the proper verification method. There are four types of verification methods: **analysis**, **inspection**, **demonstration**, and **test**. You can verify with more than one method, if necessary.

- Verification by **analysis** uses calculations or simulations to demonstrate that the requirement has been met.
- Verification by **inspection** is a physical examination, either visual or conducted through other physical means. You'll be able to observe the characteristics of your design to prove your requirement has been met.
- Verification by **demonstration** is met by performing a certain function of your system.
- Verification by **test** involves conducting a test on your system and collecting data to prove that your requirement is met.

So, how do we apply this to the sandwich we designed? (You're going to see sandwiches used as examples throughout this book) Well, let's take a step back, review our requirements, and write verifications for each. By this point in the product's lifecycle, you have designed and built a prototype or test article to begin verification.

As the lead systems engineer, you begin guiding the sandwich technicians as they build the prototype. As a team, you begin work on a configuration-controlled article in accordance with the requirement specification.

Your bread team purchases bread from the supplier. The same goes for your peanut butter and jelly teams. The teams then gather their components, and you integrate them to assemble your sandwich.

Once that's done, it's time to verify your build at the system level.

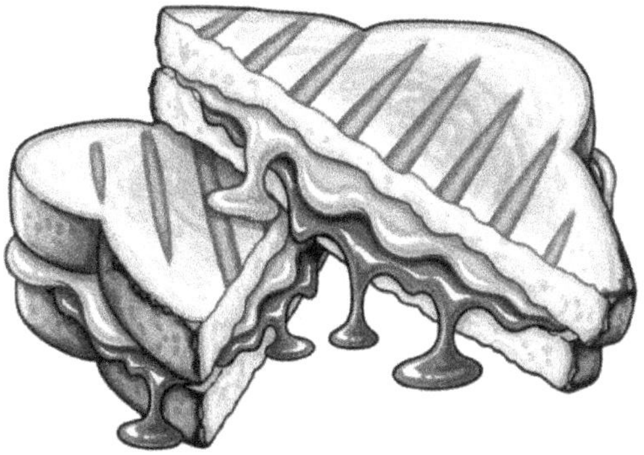

- **Requirement 1**: "The sandwich shall have two slices of whole-wheat bread."
    - **Verification method**: Inspection of the sandwich.
    - **Verification write-up**: The bread packaging label explicitly states that the bread is whole-wheat. The assembled sandwich has two slices of whole-wheat bread placed on top of one another. Therefore, by visual inspection, this sandwich has two slices of whole-wheat bread.
- **Requirement 2:** "The sandwich shall have one slice of bread spread with 0.8 oz of peanut butter."
    - **Verification method**: Test the bread by using a calibrated scale.
    - **Verification write-up**: A single slice of bread was placed on the Eliteinator weighing scale. The mass of the single slice of bread was set to zero on the scale. The engineer spread Elite peanut butter on the slice of bread and placed it on the scale. The mass of the bread slice containing the peanut butter was 0.8 oz. Thus, the sandwich has one slice spread with 0.8 oz of peanut butter.
- **Requirement 3:** "The sandwich shall have one slice of bread spread with 0.9 oz of grape jelly."
    - **Verification method**: Test the bread by using a calibrated scale.

- **Verification write-up**: A single slice of bread was placed on the Eliteinator weighing scale. The mass of the single slice of bread was zeroed out on the scale. The engineer spread Elite grape jelly on a slice of bread and placed it on the scale. The mass of the bread slice containing the peanut butter was 0.9 oz. Thus, the sandwich has one slice spread with 0.9 oz of grape jelly.

- **Requirement 4:** "The sandwich shall be assembled with the peanut butter and grape jelly facing inward."
  - **Verification method**: Demonstration.
  - **Verification write-up**: The sandwich technician took a bite of the assembled sandwich, which consisted of two slices of bread stacked on top of one another. Upon taking a bite out of this sandwich, the technician tasted the grape jelly and the peanut butter. The technician did not observe either spread on the exterior surfaces. Therefore, the sandwich was assembled with the peanut butter and grape jelly facing inward.

One thing to note is that the verification process can be entirely subjective, so it is important to ensure requirements are written with clear, measurable intent.

Challenge: Could we verify these requirements in other ways? Could we apply more rigor? How would you choose to verify Requirements 1-4?

The validation step here ultimately rests with the customer or user. In this case, if your customer eats the sandwich they asked for and confirms it meets their expectations and satisfies their hunger, that is validation of your system.

Example bad requirements:

- The sandwich shall be made properly.
- The sandwich shall contain a good amount of peanut butter and jelly.
- The sandwich shall taste great to the user.

These are bad. They are vague, unmeasurable, subjective, and not verifiable in a reasonable manner.

Example bad verifications:

- The sandwich was reviewed and appears to have been made properly. The requirement is verified.
- The sandwich was visually inspected and was determined to contain a good amount of peanut butter and jelly.
- The technician consumed the sandwich and reported that it tasted great. Ten stars. Great sandwich.

Exercise: Writing Requirements for a Sandwich

See Appendix 1 for a full Peanut Butter and Jelly Sandwich Specification.

Let's return to the sandwich analogy. Writing clear requirements is a skill that takes refinement, so practice! Write a full specification for making a grilled-cheese sandwich instead of a PB&J. Use "shall" statements to define each step. Take this a step further and verify these requirements. Did you make the sandwich as expected?

The Lifecycle of a Complex System

Let's try something a little more complex this time. For something more complex like a rocket, our need is to deliver payloads to an intended orbit reliably, efficiently, and safely. Whether the customer is a government agency, a commercial satellite company, or a research team, their input shapes the rocket's high-level requirements.

After identifying and involving stakeholders, systems engineers define the project's overarching requirements. Here are some examples (these are not meant to be actual specification numbers for a particular existing rocket design, nor comprehensive. This is just part of the thought exercise).

- Payload Capacity: The rocket shall carry up to 15,000 lbs to low Earth orbit.
- Propulsion: The rocket's engines shall generate a minimum of 4 million lbs of thrust at liftoff.
- Environmental Tolerance: The system shall operate over a temperature range of -50°C to 70°C.

These requirements, developed collaboratively by stakeholders and engineers across multiple disciplines, can provide clear targets for the development process. The systems engineers will ensure the overall program goals are specific, measurable, and unambiguous.

Break it down

With top-level requirements defined, the rocket is divided into subsystem requirement specifications. These subsystems include propulsion, guidance, structures, and payload accommodation, among others.

Here's how those requirements trace downwards:

- Propulsion System:
  - The engines shall generate 4 million lbs of thrust.
  - The propulsion system shall operate with 99% reliability with .95 confidence.
- Guidance System:
  - The software shall provide trajectory corrections within 5 milliseconds.
  - Positional accuracy shall remain within ±2 meters during flight.
- Structural System:
  - The rocket's fuselage shall endure 12 Gs.
  - The rocket chassis shall be made of stainless steel alloys.

Building and Assembling the Rocket

Once designs are finalized, manufacturing begins. Each component must meet strict specifications:

- Engines: Built with heat-resistant alloys to withstand 3,200°C.
- Fuel Tanks: Tested for micro-leaks to ensure zero loss during storage and flight.
- Fairing: Designed to protect payloads while minimizing aerodynamic drag.

During assembly, teams integrate subsystems, testing for compatibility at each step. For example, engineers might discover that the new guidance system software conflicts with the propulsion system's control software. Systems engineers can step in here to facilitate collaboration between software and propulsion teams to resolve the issue. This results in the systems engineer taking technical ownership of the integrated product.

V&V

- **Verification**: Does the rocket meet its design specifications?
    - Example: Running engine tests to confirm they generate 4 million lbs of thrust.
- **Validation**: Does the rocket fulfill stakeholder needs?
    - Example: Conducting a simulated mission to verify payload delivery accuracy.

Once the rocket is fired off, post-launch analysis provides critical insights for improving future designs:

- A review of flight data might reveal that the fuel pump vibrated at certain speeds, resulting in reduced margins and minor inefficiencies.
- Thermal analysis could suggest using different insulation materials for the fuel tanks in extreme environments.

The final deliverable is then taken and presented to the customer as a large, integrated project.

An Emergent Integration Scenario

Imagine this scenario: During integration, engineers discover that the propulsion system generates vibrations that degrade the guidance system's accuracy. Without intervention, the rocket's trajectory could deviate, jeopardizing the mission.

Your systems engineers could coordinate a solution for the anomaly:

- Vibration dampers are added to the propulsion system.
- Guidance software is updated to compensate for minor oscillations.
- Structural engineers verify that the added dampers don't compromise the chassis's integrity.

When high-level anomalies or high-stakes integration items arise, they often unfold in a nebulous interdisciplinary capacity. Systems thinking teaches us to keep all these disciplines in mind and to identify the immediate stakeholders of the issue. This is a key aspect of the integration process.

In the integration and test lifecycle, you can expect many scenarios like this. It's very rare to get it right on the first try, but that's why we design the way we do.

My Approach To Integration Problems

I love solving integration problems! It's some of the most rewarding yet complicated work you can take on as an engineer. Your most important task as an integrator is to form an educated opinion, plain and simple. It's one of the characteristics that makes a great engineer. The key to high level integration is understanding your personal style of integrating.

For instance, if you were sitting at a dinner table at a restaurant with several of your colleagues and you were responsible for placing the order, how would you approach it? What variables would you take into account? What if your colleagues don't really know what they want, but they can tell you about their dietary requirements? This could be a little daunting, but it's your responsibility to satisfy your stakeholders. This is a general basis of integration. Another thing to note is that you won't necessarily satisfy everyone with the decisions you make, but it's still important that you do form an opinion and make a decision.

These are my steps to integration problems during engineering development:

1. **Take your own cut at the problem.** Even if you don't necessarily know how you'll resolve the problem, try and take a cut to generate an early opinion. You can change this opinion based on inputs you gather, but it'll at least help you guide a conversation. In this step, you should identify requirements or components you imagine may be impacted by the technical decision.
2. **Identify your stakeholders.** Who are all your stakeholders? Not just the technical ones. Do you need to involve finance? Do you need to involve your customer?
3. **Lead conversations with your stakeholders.** Set up meetings. Pull as many of these stakeholders into a single room or forum as possible to coordinate conversations. It is crucial that you take the leadership step and establish yourself as the technical owner of this integrated product. You are the point of contact for the decisions made as a result of this effort.
4. **Form an educated opinion.** This is the fun part. Once you have an educated opinion, it's your responsibility to drive this to the finish line. Keep nurturing conversations with your stakeholders either individually or in additional integration forums. Try to be mindful of your stakeholders time and do as much due diligence on your own leading up to these conversations.
5. **Deliver the integrated decision.** Bring the integration task to a close and deliver your decision. A good decision will have risk assessment, cost impact, technical justification, concurrence from the majority (if not all) of your stakeholders, and a final approval from a review board. Bring all your receipts to this integrated decision to show that you did your homework.

Exercise: Brainstorm how you would approach the dining table integration scenario. What are some ways YOU can become a better integrator?

The How-Not-To-Crash Takeaways From Chapter 2

- If it wasn't clear by the sandwich exercise, you can design practically anything like a complex system.
- Writing targeted requirements, while potentially tedious, can make your life a lot easier by clarifying how to satisfy the end user.
- Communicate your needs across your teams to achieve a more cohesive end product.
- Integration involves forming a technical opinion.

Chapter 3: Managing Risks and Solving Problems

Every ambitious project comes with risks. From unforeseen technical challenges to human error, risks are inherent to systems engineering. You can't eliminate risks entirely through your design definition. That's impossible. Risks will continually emerge throughout the design cycle, so being proactive is instrumental in catching and mitigating them. Some risks you'll know right out the gate, but some will pop up through anomalies or coordinated discussion.

Understanding Risk in Systems Engineering

Risk is about understanding the impact of failures and the likelihood they'll occur. For example:

- Low Risk: A slight delay in the delivery of non-critical components.
- High Risk: A flaw in the power system design that could cause the lunar base to lose electricity during the lunar night.

Risks are assessed on two axes:

1. Probability: How likely is this to happen?
2. Impact: How severe will the consequences be if it does?

The combination of these factors helps prioritize risks, allowing teams to focus resources on the most critical issues. Every organization categorizes risks in its own way and has its own principles for managing them. Now let's talk about how engineers can mitigate risks.

The Risk Management Process

Risk management is a structured process that involves:

1. **Identifying Risks**:
    - Systems engineers conduct brainstorming sessions, simulations, and historical analyses to uncover potential risks.
    - Example: During early lunar base design, engineers identified the risk of dust accumulation on solar panels reducing power efficiency.
2. **Analyzing Risks**:
    - Each risk is evaluated for its probability and impact, often using tools like risk matrices or Failure Modes and Effects Analysis (FMEA).
3. **Developing Mitigation Strategies**:
    - Systems engineers propose solutions to reduce the likelihood of risks or minimize their impact.
    - Example: Adding dust-resistant coatings to solar panels and designing robotic arms to clean them regularly.
4. **Monitoring and Adjusting:**
    - Risks are continuously tracked throughout the project, with mitigation strategies updated as new information emerges.

The Fishbone Diagram

One powerful tool systems engineers use to solve problems is the **fishbone diagram** (also known as the **Ishikawa diagram**). It helps identify the root causes of an issue by categorizing potential sources and mitigating them before they arise. These sources can be referred to as *failure modes*. To identify potential failure modes that could become risks, you would hold a brainstorming session. Let's refer back to the Lifecycle V.

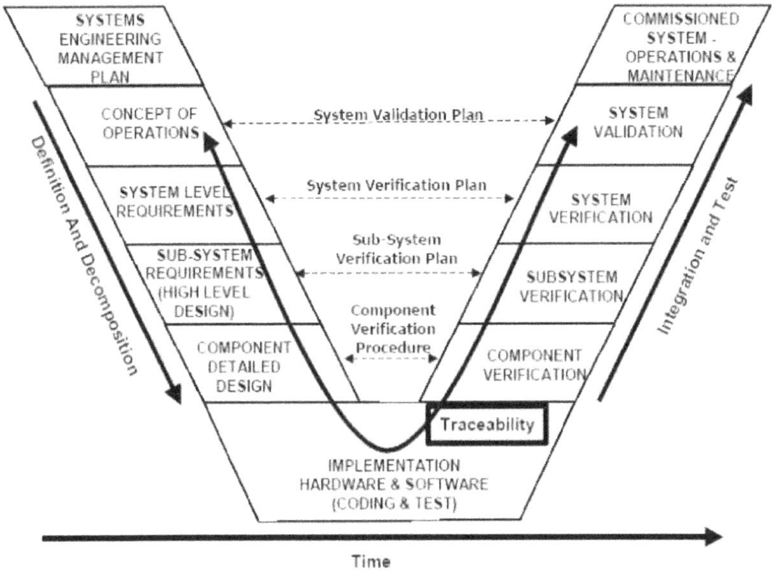

Where does risk management happen during the lifecycle? Well the answer is pretty simple… always. You can expect your mitigation techniques to mature as you progress through a product's lifecycle.

Let's talk about sandwiches again. What are some credible failure modes to a customer's consumption of a sandwich? We'll brainstorm a few together.

- The sandwich falls apart when the customer tries to eat it.
- The sandwich has the wrong ingredients.
- The sandwich bread is moldy.
- The sandwich tastes bad.
- The sandwich is soggy.

We can determine plenty more, but this is good enough to start with. Once we establish these failure modes, we can continue dissecting them. The failure modes you've determined could possibly break down into smaller failure modes. Your overarching fishbone would look like this.

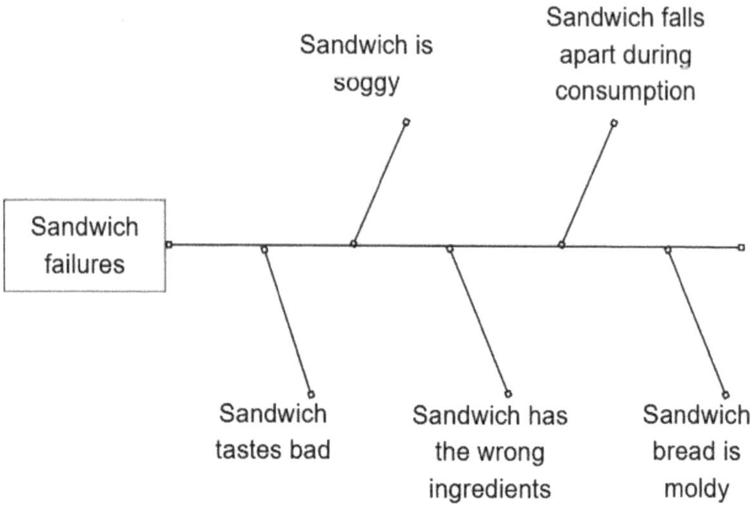

When you break it down even further, for instance: "The sandwich bread is moldy." You could have sub-causes, "The sandwich bread is not refrigerated," or "The sandwich bread was prepared in an unsanitary environment," amongst others. By working your way to the lowest cause of a failure mode, you can adequately determine the overarching failure mode's risk posture to the system. The subsequent sub-bone, in this case, would look like this:

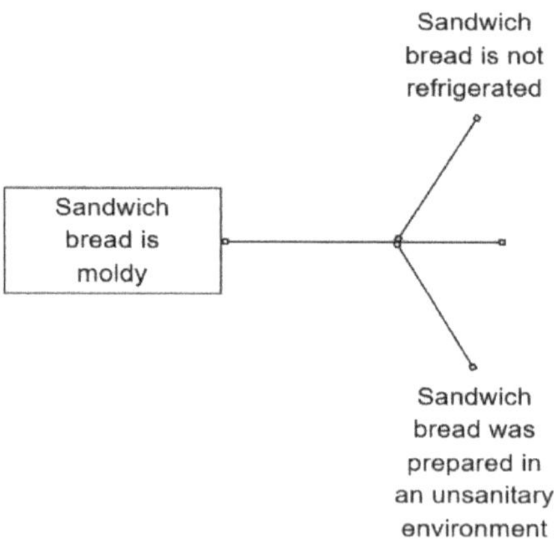

In a similar vein, as a systems engineer, you are also empowered to determine if a failure mode is excessive: “The sandwich tastes bad.” This failure mode isn’t necessarily credible because you designed this sandwich to a specification. The specification was driven by a customer's desire. The customer desires this particular sandwich to whet their appetite. Per your requirements, you’ve developed the sandwich. If the customer doesn’t like the sandwich exactly the way you’ve designed it with them in mind, that’s not necessarily a risk you can mitigate.

So let’s look at a credible one and mitigate it.

“The sandwich falls apart when the customer tries to eat it.”

1. Identify the Problem:
    - Problem: The sandwich needs to remain intact during consumption.
    - Observation: The bread tears, and the fillings slide out.
2. Analyze the Root Cause:
    - Cause and Effect on Subsystems: The bread is too soft to hold the fillings.
    - Impacts on Personnel: The assembler did not spread the peanut butter evenly, resulting in weak spots.
    - Risk Level: Medium; messy sandwiches discourage repeat consumption.
3. Resolve the Issue:
    - Analysis: Assess bread quality for structural integrity.
    - Testing: Test spreading techniques to determine optimal coverage.
    - Design: Use thicker bread and adjust filling ratios.
    - Limitations: Ensure bread thickness doesn’t compromise taste.
    - Risk Level After Mitigation: Very Low.

Note: It is possible to identify a need for an altered requirement or to develop a new requirement as a mitigation tactic. Be mindful of your specification requirements through this process. Also, be mindful of producing unnecessary mitigation steps, as this could overconstrain your project or significantly increase your development time.

Let’s look at a slightly more technical example to understand the interdisciplinary aspect of this.

Example Failure Mode: The oxygen generator in the lunar base malfunctions during testing.

Using a fishbone diagram, the team organizes potential causes into categories:

1. Subsystem Issues: A faulty sensor in the generator.
2. Personnel Impacts: A procedural error during operation.
3. Risk Levels: A mismatch between the generator’s specifications and the power system’s output poses a threat to personnel. This would be of High risk.

Mitigations include:

- Analysis: Reviewing test data to pinpoint anomalies.
- Testing: Conducting controlled experiments to reproduce the oxygen generator malfunction.
- Design: Updating the generator’s relays to better handle power fluctuations.
- Limitations: Establishing operational constraints to prevent similar issues.

By visualizing these factors, the team can systematically address the problem.

Effective risk management and problem-solving require input from multiple disciplines. This is another interdisciplinary task for systems engineers to integrate. It would’ve been imperative to bring together electrical, mechanical, and operations team members, as well as other stakeholders as needed, to understand the involved risks and the risk posture afterward. In this scenario, after the team determines the mitigations and the credibility of the failure mode, they can indicate that it's been mitigated to a Low risk category.

Exercise 1:

Determine mitigations for each of the remaining sub-bones, if they are credible.

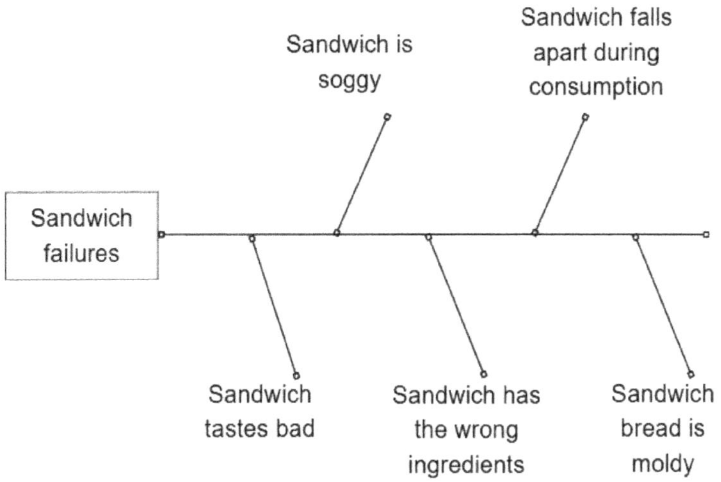

Exercise 2:

Let's revisit the kitchen from Chapter 2. Imagine the team identifies a risk: excessive moisture and smoke from the kitchen make it really hard to work in and way too hot.

1. Write three potential root causes using the fishbone categories:
    - Subsystem Issues: Inefficient kitchen ventilation.
    - Personnel Impacts: Kitchen staff is getting way too hot.
    - Risk Levels: The HVAC system is undersized for the base's needs.
2. Develop a mitigation plan:
    - Add a high-efficiency dehumidifier to the kitchen.
    - Update crew training to emphasize moisture control.

    - Increase the HVAC system's capacity to handle additional loads.

3. How would you validate these mitigations?

The How-Not-To-Crash Takeaways From Chapter 3

- Risk management involves considerable brainstorming.
- Work your way down to the lowest level of a failure mode before determining the overarching risk posture.
- It is okay to determine if a failure mode is either too unrealistic or too arbitrary.

Chapter 4: Evaluating Trade-offs

The Balancing Act of Systems Engineering

Imagine you're designing a bridge. It has to support heavy traffic without being so over-engineered that it bankrupts the city. It also needs to be built quickly, yet cutting corners could lead to catastrophic failure. Balancing these competing priorities (cost, schedule, performance, and risk) is the heart of systems thinking.

This chapter explores how engineers evaluate trade-offs, prioritize objectives, and manage risks to deliver solutions that satisfy stakeholders' needs while ensuring safety and reliability. You'll often hear these tasks referred to as "trade studies."

The Trade Trio

Every project involves three interdependent factors:

1. **Performance:** How well the system fulfills its requirements.
2. **Cost:** The financial resources required to develop, build, and maintain the system.
3. **Schedule:** The time needed to develop and deliver the system.

Adjusting one factor impacts the others:

- Improving performance often increases costs or delays schedules.
- Reducing costs might compromise performance or require more time.
- Accelerating the schedule could lead to increased costs or reduced quality.

Trade-offs in Risk Mitigation

Mitigating risks often involves trade-offs, as solutions can introduce new challenges. For example, let’s take rocket insulation: one benefit is that it prevents propellant freezing. One drawback is that it adds weight, slightly reducing payload capacity.

You have to take these factors into consideration when developing a trade study and conducting a trade-off analysis.

Exercise: Evaluating Trade-offs in Sandwich Design

You’re tasked with creating the ultimate PB&J sandwich. Here's the scenario:

Problem: The sandwich needs to balance taste, structural integrity, and ease of preparation.

1. **A Performance Requirement From The Customer:** The sandwich shall hold together without spilling and have a balanced flavor.
2. Constraints:
    - Cost: Ingredients should not exceed $5.
    - Schedule: The sandwich needs to be prepared within 10 minutes.

Trade-offs:

- **Option 1:** Use thick peanut butter and jelly layers.
    - **Benefit:** Enhances flavor.
    - **Drawback:** Increases the risk of spillage and requires more time for spreading.
- **Option 2:** Toast the bread.
    - **Benefit:** Improves structural integrity and adds texture.
    - **Drawback:** Adds preparation time and slightly alters flavor.

- **Option 3:** Use honey instead of jelly.
  - **Benefit:** Reduces spillage risk and adds a unique flavor.
  - **Drawback:** Alters the classic PB&J taste and slightly increases cost.

Which option would you choose, and why?

Here is a decision matrix comparing the three sandwich design options, listing their impacts on flavor, cost, and preparation time.

Option	Flavor	Cost	Ease of Prep	Structural Integrity
Thick PB&J Layers	High	Medium	Medium	Low
Toast the Bread	Medium	Low	Low	High
Honey Instead of Jelly	Medium	Medium	High	Medium

The How-Not-To-Crash Takeaways From Chapter 4

- Trade-offs are inevitable in complex projects; systems engineers ensure they're made strategically.
- Risk management is essential to prevent small issues from escalating into catastrophic failures.
- Systems engineers balance technical, financial, and operational priorities to ensure projects meet both immediate and long-term goals.

Chapter 5: Iterations and Upgrades

Everything we use undergoes iterations and upgrades (some more drastic than others). Consider how iPhones have evolved over the past ten years, how your television has changed since childhood, or the differences between a new 2026 Toyota Corolla and one from 1997. Designs tend to improve over time. There is a slightly more cynical take on future development that I will discuss further in the next chapter. But we'll approach this concept more methodically and assume proper design optimization.

Systems engineering is an iterative process where early decisions inform later stages, and insights gained later can refine initial goals. Similar to perfecting a recipe, developing complex systems requires cycles of development, testing, and refinement. When you upgrade or iterate on a design, the development process is expected to be much faster.

We'll use a city-wide public transportation system as an example to illustrate this process. Public transit systems routinely undergo upgrades to accommodate traffic flow and underlying local projects. We can start with a base system design and upgrade it from there for this exercise.

Iteration in the System Lifecycle

The development lifecycle of a system is usually a loop. Here's how iteration integrates with the classic systems engineering lifecycle:

Concept Development

Iterations or upgrades begin back at the conceptual level on our favorite Lifecycle V, once more. At this point, you would develop your systems engineering and management plan before moving on to the concept of operations. The initial idea for a city's transportation system might be as simple as "The system needs to reduce average commute times by 30%."

Early feasibility studies from the first deployment might reveal conflicting factors, such as limited infrastructure or budget constraints. The concept evolves through discussions with stakeholders, shaping the high-level design upgrade into a realistic, achievable plan.

Many engineering firms use the terms **block upgrade**, **modification**, or **revision** to distinguish design iterations over phases.

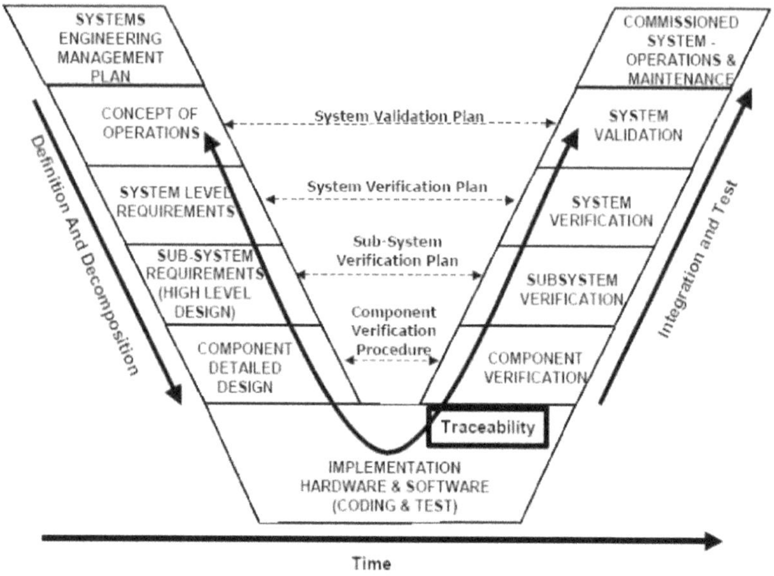

Example Iteration:

- **Initial Concept:** A network of autonomous electric buses.
- **Refinement:** After stakeholder input, the focus shifts to making it easier for the buses to get around. There's a need for more charging stations for the buses alongside dedicated bus lanes.

Project Definition and Requirements Development

As the concept evolves, so do the specifications. Iteration ensures requirements remain feasible and aligned with the overarching goals. Initial designs are rarely perfect, and prototypes are built to test assumptions. Each iteration may uncover or create new flaws.

Example:

- **Design Block 1:** Buses struggle with steep inclines in hilly parts of the city.
- **Design Block 2:** Engineers adjust motor specifications to handle hills and add regenerative braking to improve efficiency, but find that an additional power draw from the batteries is required.

While you're upgrading or improving a design, it's important to understand that re-verification of EVERYTHING isn't always in your best interest. Doing that can greatly increase a project's cost and scope. If you know that a subsystem in your system already works and there are no changes, you don't need to pursue reverifying requirements, nor do you need to write new ones.

This process is called **requirements allocation**. In a case where you have infinite time and infinite money, you could theoretically reverify everything, but the real world doesn't typically work that way.

For instance, if one of the fleet upgrades being made is longer-lasting batteries, during your requirements allocation *scrub* of the specification, you would identify what is impacted by the battery upgrade that may drive reverification. The best practice in a situation like this is to ensure that you have a rationale for why you didn't choose to allocate a requirement to be reverified.

Note: Your verification method may also evolve as further design upgrades are made.

While you scrub initial requirements for relevance, you may also need to tweak them.

- **Initial Requirement:** "The bus fleet shall have a range of 500 km on a single charge."
- **Refined Requirements:** "The bus fleet shall have a range of 350 km on a single charge."

"Dedicated bus fleet charging stations shall be placed within 5 km of each other."

Exercise 1:

Your electric bus fleet is mandated to increase battery capacity to 450 kW·h to accommodate a chassis update. This update would make the buses safer to operate; however, they would be heavier.

In the list of requirements here, which ones would you initially pick out to be reverified for your bus fleet? How would you modify any of these requirements?

- The bus chassis shall be comprised of high-strength steel.
- The bus shall carry up to 100 passengers.
- The bus battery system shall generate 300 kWh of energy.
- The bus shall be equipped with a GPS navigation system.

Exercise 2:

For the requirements that you've selected to allocate above, how would you approach verifying them?

Integration, Deployment, and Operations

The remainder of the upgrade's lifecycle moves as shown in Chapter 2 through the Lifecycle V.

Systems continue to evolve even after deployment. Fortunately, but also unfortunately, there's no better way to discover a design flaw than to find one out in the wild.

In this day and age, with the rigor of our systems engineering processes, it's a lot safer. Operational feedback provides valuable data for future updates, whether to fix bugs or enhance performance.

Some Systems Engineering Best Practices for Iteration

- Start targeted and then upgrade. Early prototypes need to answer specific questions. Use these insights to scale up gradually. Develop, iterate, and then upgrade.
- Leverage stakeholder feedback. Stakeholders, including end users, provide invaluable perspectives. Build mechanisms to collect and act on their feedback throughout the lifecycle.
- Embrace failures as learning opportunities. Failures in early design iterations are far less costly than failures in deployment. Encourage a culture where failing hard during development is seen as an opportunity to learn. Don't be like Boeing with the 737 MAX.
- Document everything. Keep detailed records including decisions made, changes implemented, and lessons learned. This ensures transparency and helps future iterations build on past successes.

The How-Not-To-Crash Takeaways From Chapter 5

- Iteration drives innovation and refinement. It allows teams to test ideas, identify flaws, and implement improvements.
- Iteration is where development gets tangibly faster.
- Feedback loops ensure systems meet real-world needs. Stakeholders, testing data, and operational feedback are invaluable.
- Documentation and transparency are critical. They ensure that each iteration builds on prior knowledge.
- Failure is an essential part of success. Early missteps pave the way for better outcomes in later stages.

Chapter 6: The Future of Engineering and Problems We've Yet To Solve

A Blurb On Model-Based Systems Engineering

In today's fast-paced development landscape, one of the most pressing challenges we face is the quest for speed. As projects grow in complexity and timelines shrink, the demand for faster, more efficient development processes becomes more critical. Teams are constantly searching for innovative strategies to accelerate production without compromising quality or overshooting budgets.

This urgency to evolve invites us to rethink traditional methodologies and embrace a culture of agility and collaboration. By leveraging upcoming tools and fostering an environment that encourages creative problem-solving, we can streamline workflows and enhance productivity. Ultimately, the goal is to create a development approach that is swift and sustainable. This thinking spawned the concept of **model-based systems engineering (MBSE).**

MBSE is a relatively new approach to viewing a system at a high level. Per the namesake, it's a way to visualize systems as models that can be broken down through visual interfaces. Instead of treating requirements, design, and verification as separate pieces, they are all connected within a unified model. When something changes, you can see the impact across the entire system instead of discovering it later during integration. This would drastically reduce risk during engineering development and cut down on tedious overhead causing development slogs.

You don't need to master this right now (this isn't really the point of the book). The goal is to recognize that this is where engineering is heading. The common graphical modeling language utilized for MBSE is called SysML. As systems continue to grow in complexity, the ability to visualize connected models to understand relationships across a system will become increasingly valuable.

A Blurb On Innovation (Especially as Artificial Intelligence Enters The Scene)

Curiosity is what drives innovation. With artificial intelligence taking the forefront, we're put in a new era of innovation. Some theorize that artificial intelligence will take over everything! Some theorize that it'll become the new tool we use, allowing us to be more efficient at developing things, akin to a calculator.

My personal view is that it will be a balance of both. Automation and autonomous functionality already exist in many areas of engineering development, such as autopilot systems, where machines are designed to operate with minimal human input. In some industries, automation has replaced roles, like in warehouses, where robots can perform the work of multiple technicians.

At the same time, there is another side to consider. If artificial intelligence becomes more integrated into development, it could accelerate execution. People with strong ideas may be able to move faster, design more efficiently, and bring concepts to life with fewer barriers. This is where you come in! The main question is how the nature of work evolves, and how engineers adapt to this big upcoming shift. Well… nobody really knows, but if you are reading this, you are likely part of the future. Never underestimate the value of your humanity.

Humanity is ultimately the strongest determinant of a design's value. No matter how advanced or technically impressive a system may be, its true worth is defined by how well it serves people. Engineering development is all about solving real problems for real users, so the most impactful solutions are those that keep people at the center and deliver meaningful value through a product. From an optimistic standpoint, I say the key to innovation is humanity.

The future of engineering will be shaped by how well we adapt to new tools, our ability to think in systems, and our continued commitment to placing human value at the center of what we build. The engineers who embrace this shift will be the ones who drive meaningful impact.

Problems That Still Need To Be Solved

A development principle that I am very critical of is the same philosophy that made Boeing cut corners. This is the one that can lead to a more cynical take on upgrades. Essentially, with so much of development and innovation being tied to increasing shareholder profits, there is a tendency for companies to make things cheaper to develop rather than simply making higher quality things. We may lovingly refer to this principle as **"enshittification"** for lack of better terminology.

This is the process of things being produced with declining quality. You can see enshittication all around you: your chain-restaurant orders, modern cars, electronics, rushed video game developments, apps, etc. Frankly, this is really unfortunate. This kind of decision-making tends to come from executives that are out of touch with societal development needs and simply want to raise profits and secure larger bonuses. This is an abhorrent practice and should be discouraged. It's the reason things are getting more and more expensive and seem to be made a lot lower quality. Don't let this happen. Lowering your production costs is a good thing, but lowering your development standards at that expense is not. Make better stuff and please don't "enshittify" your development.

Now with my blurb on enshittification in mind, one of the most important issues in the industry right now is the need to develop faster (ideally without making things shittier). This is a crucial need for prospective engineers with systems engineering knowledge, which is why I've hammered on the element of increasing development efficiency throughout this book.

Now, as a specific example, energy is a field that consistently needs innovation. One of the biggest issues we're looking to solve worldwide is how computers and large language models consume energy. You'll notice that there are a million start-ups consistently getting funded for AI development, energy solutions, and satellite solutions. Given the extreme nature of competition in these industries internationally, there's a push to develop fast or fail hard.

A list of Top 100 Rising European Startups in 2026 - ohn Pessara Editorial Copywriter @VivaTech

These are gaps that you, as an engineer, can solve. There are no man-made problems too big that can't be solved with a man-made solution. All it takes is a strong value proposition and ingenuity.

In the next chapter, we'll talk about how you can take the things you already have in your repertoire, combined with systems engineering knowledge, to begin making a difference by positioning yourself as a highly valuable candidate and a capable engineer.

The How-Not-To-Crash Takeaways From Chapter 6

- Engineering is shifting toward model-based approaches. Connected models and traceability are becoming the standard for faster development.
- Avoid enshittification as best as you can. It's making the world a worse place. It's not good for anybody. Seriously.
- Energy is one of the largest sectors demanding innovation.
- The future will reward adaptability, particularly with the emergence of artificial intelligence.
- Your humanity is your advantage.

Chapter 7: How to Build a Solid Resume and Nail an Interview

At this point, you have a working understanding of systems thinking. You can own that and apply it to building your resume! You can write a line indicating that you are familiar with systems engineering. The purpose of this book is to ultimately create better engineers and improve your career trajectory. While we can say a lot more about systems engineering as a technical function, I want to bring the focus back to establishing yourself as a capable and effective engineer. Principles of systems thinking can be used to plug-and-play into any engineering discipline.

If there's any one thing I want you to take away from this chapter, it is that YOU HAVE VALUE, no matter where you are in life! Even if you feel as though you're falling behind, you have value. Firstly, let's dispel a few myths:

Myth 1: "Your college education is the only way to achieve a successful engineering career."

You do not need to have attended a particular college over another to maintain a competitive edge in the job market. It can help with networking and with opportunities to attend a bigger school, sure. Ultimately, the way to achieve success as an engineer is to be the best candidate, no matter your background. Sometimes, to get conversations riled up, I'll drop a blanket statement that the modern college system isn't helpful for creating good engineers, to draw some attention and dispel this myth in depth. What is true is that those formative years give you the neuroplasticity it takes to develop into a formidable engineer rather than teach you concepts that you'll remember forever. The reality is that the years you spend in your education can be used to find opportunities that you won't be able to find in other places, such as career fairs, networking events with industry professionals, research opportunities, and clubs. That is where the true value of a college degree lies. YOU attend the interview, not your degree. Anyone with a desire to achieve and learn can fill that role. You learn how to be an engineer in the role, not before.

Myth 2: "Pursuing a Master's degree gives you a career boost you wouldn't get otherwise."

I think there was a time where this was very true. Given how saturated the market is right now and the record number of students pursuing a university education with a Master's degree, there's an itch to find candidates with genuine experience. Now, there IS value in a Master's degree and that value comes from a student's pursuit of a particular discipline. This is a great opportunity to pursue research work or learn about a subject matter in detail such that you can take away to use within your career. The "career boost" comes from HOW you pursue your Master's degree, not IF.

Myth 3: "Pursuing a trade is less valuable than pursuing a university education."

This is also false. These don't have to be mutually exclusive things either. It's as simple as this: if you had an issue with your car, would you rather contact a mechanic or a mechanical engineer? You'd likely call the mechanic, and that's because of the field experience. If you wanted to understand the phenomenon behind the issue a mechanic found in your car, you'd likely contact a mechanical engineer. Having the technical depth and the neuroplasticity that come from a college degree is still valuable; however, a tradesman can maintain practical hands-on knowledge that is just as valuable. These things go hand in hand when it comes to engineering development. Remember, think in systems. Every service, process, and component plays a role in the final outcome of developing a solid product.

Myth 4: "Artificial intelligence is taking over, and there's no hope getting through to a real recruiter."

The job market is incredibly competitive, and with the introduction of AI, it can get really disheartening. I promise you, there are still humans being hired for those roles you see posted online. Even if the market is competitive or infested with AI application filters, the need for good engineers will always exist. I believe anyone can be a good engineer. The key to succeeding and standing out is your grasp of systems thinking.

Recognizing a system's functionality and how you can contribute to it is one of the BEST value-adds you can bring as an engineering candidate. This applies whether you're chasing upward mobility, switching disciplines, or even just getting your feet wet in the industry. A lot of what you learn in an engineering role, you really do learn while you're on the job.

Building a Solid Resume and Using Systems Thinking

It's genuinely a really difficult thing to reflect on your skills and experience to sell yourself on a resume. Your resume should be a living document so that every time you open it, you can find SOMETHING to update. The trick to building a solid resume is understanding what you can talk to a recruiter about. It cannot be understated: DO NOT LIE on your resume. You don't need to embellish details to be a solid candidate. You just need to know how your experiences and skills will solve the recruiter's problem. One more thing, it really helps to have a peer as a resource to review your resume.

First things first, you need to understand that when a hiring manager or recruiter is looking to fill a role, they're seeking a candidate to meet a very specific need within their organization. Another way to look at it is that they'd like to improve a certain element of their *system*. For that reason, your recruiter will look for specific skills to ensure compatibility and onboard a prospective employee as quickly as possible. With that being said, let's talk about how to stand out.

Take Advantage of Your Early Career

So, how do we do it? You need to accept that no matter what you write on your resume, you are underselling yourself. Your early career can be a goldmine of opportunities. You'll generally have exposure to a variety of different subjects and interest fields earlier in your career to refine your placement as a more specialized engineer as you grow. Your university education and extracurriculars can also be considered part of your early career when designing a solid resume.

A lot of hiring managers (or their AI filtering system) look for keywords in a candidate's experience history to latch onto and assess their background. This is where you can leverage your university specialization and projects you've completed. People tend to forget that what you do in school can genuinely count for experience! You go to engineering school to gain the neuroplasticity to be a functional engineer.

For example, if you are an electrical engineering student and, for one of your school projects, you developed an Analog-to-Digital converter (ADC) circuit, this is still circuit design experience! Take ownership of that in your resume. In the same way, take ownership of courses you've gone through and performed well in. It all counts.

For example, instead of saying: "Designed an ADC circuit for a class project."

You can say: "Designed and tested an analog-to-digital converter circuit to satisfy a theoretical user story."

With a simple tweak like that, the project already sounds like it has more of an impact.

Beyond just being a student, you can leverage your experience from prior roles or things you've been a part of. Ex) customer service, sales, part-time work, tutoring, volunteering, etc. We'll continue to break down how all of this can build your professional resume.

Own Your Skills!

Your resume isn't a place to be humble, so don't downplay what you're capable of. Don't be shy about owning both your hard skills AND your soft skills.

Your skills run deeper than you realize. My philosophy is simple: if you have worked with something and can pick it back up and improve on it, that is a skill. Even if you are new to it, it still counts because there is value in exposure. There is value in showing that you can learn and build on the things you learn. If you have used a tool, written code, built a circuit, led a discussion, or contributed to a project, these are things you bring to the table. Call it out clearly. You may indicate you're new to the skill, sure, but it's still a skill nevertheless.

It's easier demonstrated, so here are some broad examples that you can use to directly apply similar logic to your resume.

Example 1: A solo class project involving the design of a system.

Some takeaway skills:

- Applying engineering fundamentals to a real design problem

- Debugging and troubleshooting when the design does not work as expected
- Using tools such as CAD, MATLAB, LTspice, or Python to model your design or test behavior.
- Breaking down a problem into smaller, manageable parts.
- Iterating on a design based on test results or failures.
- Documenting design decisions and results clearly.

Example 2: A group class project.

Some takeaway skills:

- Technical communication across team members (bonus points if it's cross-disciplinary)
- Task coordination and time management
- Conflict resolution and decision making
- Accountability and ownership of deliverables
- Exposure to different technical approaches
- Technical presentation experience
- Leadership

Example 3: Lab work.

Some takeaway skills:

- Test execution and data collection
- Following structured procedures
- Analyzing and interpreting results
- Identifying anomalies and root causes
- Attention to detail
- Writing clear and concise lab reports

Example 4: Engineering Coursework

Some takeaway skills:

- Strong foundation in core subjects such as circuits, thermodynamics, or mechanics
- Ability to solve structured technical problems under time constraints
- Applying mathematical concepts such as calculus and differential equations to real problems
- Familiarity with engineering tools and software used in coursework
- Reading and interpreting technical problems and requirements
- Developing discipline and problem-solving habits

Example 5: Internship, Co-op, or Early Career Engineering Role

Some takeaway skills:

- Contributing to a larger engineering system with many stakeholders
- Understanding how individual work impacts upstream and downstream components
- Applying systems thinking to identify dependencies and potential issues early
- Communicating technical progress, risks, and decisions to team members or leadership
- Taking ownership of tasks from assignment through completion
- Working within real-world constraints such as schedule, cost, and requirements
- Collaborating across disciplines such as electrical, mechanical, software, or systems teams
- Supporting testing, integration, or validation activities

Example 6: Misc Non-Technical Experience

Some takeaway skills:

- Contributing to a team setting forth for a particular goal.
- Achieving something as part of a volunteer group.
- Performing a customer-facing service in any role. Ex) front desk worker, waiter, host or hostess, wilderness guide, farmhand, etc.
- Customer relations.
- Business development or sales exposure.
- Business sense and intuition.
- Performing a service that keeps a system in place.
- Leadership experience.
- Discipline from being part of a sports group or other type of rigorous, time-demanding activity.

When you break down your skills this way, you show that you understand what a hiring manager is actually looking for and that you can step into the role they need to fill.

Ask For Help

Asking someone with more or similar professional experience to review your resume is crucial. Building a resume can be a really tough exercise, so it's super beneficial to simply ask your peers or mentors if they can help you out by asking you questions about yourself. I've personally found value in having my friends ask me detailed questions about my professional background to get the gears churning. It helps to take notes during this process as well.

Something else I've found personally helpful is having someone who just knows you well as a person review your resume. These people are the best at determining whether you are presenting yourself authentically.

Get Well-Rounded

Being well-rounded in your life can lead to being a better engineer. Being a better engineer allows you to develop more as a person and bring value to yourself. In a sense, this is systems thinking in a broad manner.

Routines, mindfulness, continuously learning, and accepting that you have plenty of room to grow. Really, you can start by building a record of discipline. That record of discipline shows up in how you approach problems, communicate, and handle setbacks. You build reliability in your work by establishing consistency in your daily habits, which, over time, builds trust. Then trust, in turn, becomes responsibility, and responsibility becomes career progression. People start to rely on you because you follow through, stay sharp, and continue improving.

Even small things like how you present yourself matter. Research on *enclothed cognition* by Hajo Adam and Adam D. Galinsky (2012) shows that what you wear can influence both how others perceive you and how you perform. Individuals dressed in more professional attire were associated with greater competence, attention, and authority. It's a subtle factor that contributes to how seriously you are taken in a professional environment. Practice good hygiene and take care with the way you present yourself.

Reflect on the soft skills I mentioned in Chapter 1 to identify other ways to stay well-rounded. The most important thing to understand is that you ALWAYS have room to evolve. The best way to figure out how is by remaining true to yourself, which ultimately ties into nailing an engineering interview.

How Not to Crash In An Engineering Interview

The point of all this is to ensure that you are fully prepared for an engineering interview and you secure a strong upward career trajectory. An interview is a two-way street. It's like dating! You are interviewing your recruiter as much as they are interviewing you. Remember, you are a VALUABLE asset, and at the end of the day. At the end of the day, you've got to be happy with your work and assess if the company is a place you'd be willing to put in 8+ hours daily.

There will likely be several stages of interviews. There's the initial phone screening, the onsite interview, and potentially an exam or a project, depending on the industry, give or take a few stages based on the rigor the hiring team is looking to use. You must be prepared for all of these things. The whole concept of systems thinking comes to fruition here. When interviewing, your mission as a valuable engineer candidate is to figure out what sort of role you'll play in the bigger picture as clearly as you can manage from the recruiter. It's not uncommon to take an interview and be sold a whole "dog and pony show."

Here are some steps you can take while leveraging systems thinking:

1. **Guide the interview YOURSELF.** Be prepared to ask very relevant questions. A benchmark I like to set is to ensure that I am the one leading the conversation and asking the questions. The more strategic questions you can ask your recruiter, the better your interview will go. Any time you ask these questions, be sure to find a way to tie them to your resume. Also, take notes and learn about your recruiter. Here are a few good questions:
    a. "What are some roadblocks you are seeing in your development?"
    b. "What background are you looking for from me to help you solve these roadblocks?"
    c. "Where in the development lifecycle will I be supporting?"
    d. "How does your team currently make engineering decisions?"
    e. "Who would I work with most closely on a day-to-day?"
2. **Treat your resume like a cheat sheet.** One of the best ways to prepare for an interview is to know your resume inside and out. The whole point of a resume is to help you tell your story through your interview. Your recruiter will almost always have a copy of your resume for their reference, so it'll be easy for them to follow along with you. Once again, do NOT lie on your resume. If you're put through a project or an exam, you will need to rely on your technical depth to execute on that for the hiring team.
3. **Implement systems engineering and systems thinking, and SOLVE their problem.** Think about the larger system. What is the larger system that you're interviewing to work in? Assert yourself as the candidate who can solve the recruiter's problems, as if you were already in the role. Ask your day one technical questions now. What subsystem will you be a part of? What approach do they follow for risk management? How much

ownership will you take of this subsystem? How closely do their interdisciplinary teams work together? How would you handle an integration task? If your hiring manager tells you what their problem is, offer up a broad solution tailored to you. Make it clear that you understand the overarching picture.

For example, if a hiring manager tells you that their team is struggling during integration because different subsystems are not aligning properly, this is your moment. Instead of just acknowledging it, you can step in and say something along the lines of, “I’d want to understand where the disconnects are between the teams. How are they currently interfacing, and how does this play into the larger system? This is something I can solve with my background in ‘xyz’.”

4. **Stay hydrated.** Seriously. Stay hydrated. You’re going to be talking a lot. This’ll help you stay sharp as a tack.
5. **Know your value.** Do not compromise on your non-negotiables. Know your values and don’t let yourself be sold into something you don’t agree with. You can present your personal ethos to your interviewer through the quality of your questions. My personal take is that you should also know the value of your time and avoid subjecting yourself to interview processes that are longer than four or five stages. Also, there’s a scummy quality some startups will take advantage of, where they’ll have a candidate perform labor for free for a set period before determining whether to hire. Don’t fall for this.

The How-Not-To-Crash Takeaways From Chapter 7

- You have value no matter where you are. Don't fall for any myths that tell you otherwise.
- Your resume is a living document that will always evolve.
- Don't be humble on your resume. Don't lie on your resume. Reflect deeply on your background. Ask for help.
- Apply systems thinking and systems engineering philosophies throughout your resume and in your interview.
- Recognizing a system's functionality and how you can contribute to it is one of the BEST value-adds you can bring as an engineering candidate. This applies whether you're chasing upward mobility or switching disciplines.
- Drive the interview as an interviewee.
- Use your resume like a tool to follow along in a conversation with your recruiter.

Wrapping It All Up

At this point in this book, you should have a solid understanding of systems thinking. When you think in systems, you can quickly piece together what is being accomplished and how you can jump in and make yourself a highly capable engineer. Systems thinking and systems engineering are genuinely the way we, as engineers and big-picture thinkers, can make a big difference.

As we push for a safer, more efficient future, keep all these lessons in mind. Use this toolkit to progress tremendously in your career.

Per aspera ad astra!

Your friend,
Sai Dharani, an engineer.

See Appendix 2 for details regarding The Elite Engineer Blueprint program.

APPENDIX 1

Peanut Butter and Jelly Sandwich System Specification

Reqt 1. The sandwich shall consist of two slices of whole-wheat bread.

Reqt 2. The sandwich shall contain 0.8 oz ±0.05 oz of peanut butter.

Reqt 3. The sandwich shall contain 0.9 oz ±0.05 oz of grape jelly.

Reqt 4. The peanut butter and jelly shall be applied to opposing inner-facing surfaces.

Reqt 5. The sandwich shall be assembled with spreads facing inward.

Reqt 6. The sandwich edges shall be aligned within ±0.25 inches.

Reqt 7. The sandwich shall be safe for human consumption.

Bread Subsystem Specification

Reqt 1. The bread shall be whole-wheat.

Reqt 2. Each slice shall measure between 4.0 and 5.0 inches in width and length.

Reqt 3. Each slice shall have a thickness between 0.4 and 0.8 inches.

Reqt 4 The bread shall be free of visible mold, contamination, or damage.

Reqt 5. The bread shall be stored at kitchen temperature 65°F–75°F prior to assembly.

APPENDIX 2

The Elite Engineer Blueprint is a 6-week 1-on-1 program I've developed to help engineers and aspiring engineers accelerate their careers and land high-paying engineering roles that set them up for success. The program is tailored with three main pillars:

1. **Systems Thinking**: The content of this book is tailored to the individual.
2. **Executive Communication**: Communicating technical information effectively in executive and high-stakes environments.
3. **Ownership**: Learning how to take ownership of your system, subsystem, or "wheelhouse" in your role in order to achieve measurable success

I've put together a program through which I mentor early-career and aspiring engineers, working with clients individually to play a small role in helping them achieve their goals.

You can find more information about this program at theeliteengineer.com/blueprint.

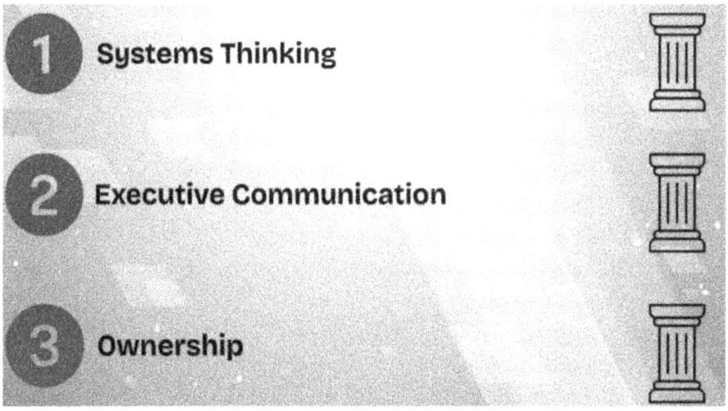

APPENDIX 3

Exercise: Try designing a lunar base!

In this era, the moon is a hot commodity sought after by many large companies and countries. The unfortunate reality is that designing a lunar base is a complex, ambitious endeavor requiring careful coordination of numerous subsystems, detailed planning, and innovative problem-solving. For fun, let's take a crack at it ourselves in a fictional scenario. This chapter will walk through the high-level lifecycle of designing a lunar base, showcasing the systems engineering process in action. Let's make this an approachable task. For this exercise, you can even use LEGOs or similar building blocks to come up with models.

Why a Lunar Base?

The Moon has captivated humanity for centuries. The Apollo missions proved it was possible to land on the Moon, and now the Artemis program aims to go even further by establishing a permanent presence there. But why build a lunar base? (Besides the fact that it'd be super cool)

- **Scientific Research:** The Moon offers a unique environment for studying planetary formation and conducting astronomy free from Earth's atmospheric interference.
- **Resource Utilization:** Water ice found in the Moon's polar regions can be used for drinking water, oxygen, and even rocket fuel.
- **Gateway to Mars:** A lunar base serves as a proving ground for technologies and systems needed for Mars exploration.

The International Lunar Coalition (ILC), a coalition of governments and private companies, has issued a call for designs for a sustainable lunar base.

They ask for a base that can support 20 astronauts for six months, using local lunar resources and pioneering new technologies.

Phase 1: Define the Requirements

The first step in any systems engineering project is to define the requirements. These are the foundations for all subsequent work.

High-Level Requirements from ILC

Habitability:

- The base shall accommodate 20 astronauts for at least 180 days.
- The habitat module shall maintain a stable internal temperature of 68°F ± 4°F.
- The life support system shall recycle at least 95% of water and provide breathable air with 21% oxygen.

Sustainability:

- The base shall utilize lunar resources for construction and support systems.
- The power system shall generate at least 67 horsepower (50 kW) during lunar days and maintain 13 horsepower (10 kW) during lunar nights, using backup systems.

Expandability:

- The base shall include modular ports for adding new habitat, research, or storage units.

Safety:

- The base shall withstand micrometeoroid impacts and provide adequate radiation shielding for the crew.

These requirements are compiled into a specification document that serves as the project's blueprint.

Phase 2: Breaking Down the System

A lunar base is composed of numerous interdependent subsystems, each with its own set of requirements:

Habitat Module:

- Shall provide sleeping quarters, a kitchen, and common areas.
- Shall include radiation shielding made from lunar regolith.

Life Support System:

- Shall recycle air and water with an efficiency of at least 95%.
- Shall maintain humidity levels between 40% and 60%.

Energy System:

- Shall consist of solar panels and battery storage to provide uninterrupted power.
- Shall include backup nuclear generators for emergencies.

Resource Utilization System:

- Shall include equipment for mining and processing water ice.
- Shall extract oxygen from regolith for breathable air.

Each subsystem is assigned to an interdisciplinary team of specialists.

Challenge: How would you divvy up your systems engineers? Can you develop more subsystems? How would you break down the subsystems?

Phase 3: Materials and Sourcing

A project of this scale requires materials sourced both from Earth and the Moon:

Earth-Based Materials: Advanced electronics, life support systems, and some structural components are manufactured on Earth and transported to the Moon.

Resource Utilization RU: Lunar regolith is used to construct radiation shields, and water ice is harvested and processed for drinking, oxygen, and hydrogen fuel.

Challenge: How would you isolate and procure materials? Who would you give ownership of these processes?

Phase 4: Integration Challenges

As the subsystems are developed, they must be integrated into a cohesive whole. This is where systems engineering shines.

Challenge:

During integration, the team discovers that solar panels generate inconsistent power due to the accumulation of lunar dust. This causes fluctuations in the life support system.

How would a systems engineer be responsible for working with the responsible engineers to develop a solution?

Phase 5: Verification and Validation

Before deployment, every subsystem undergoes rigorous testing:

Challenge: How would you validate your subsystems?

Phase 6: Delivering the Final Product

The final deliverable to ILC includes:

- **Specification Document:** A comprehensive list of requirements and how they were met.
- **Integration Report:** Details how all subsystems work together.

- **Validation Summary:** Summarizes test results and demonstrates system readiness.
- **Visual Presentations:** 3D models and videos illustrating the base design.
- **Operational Plan:** Step-by-step instructions for constructing and operating the base.

Challenge: Compile the theory! Look at your development lifecycle.

Exercise: Let's keep the lunar base development going!

You are tasked with designing a kitchen for the lunar base. Using the principles outlined in this book:

1. Write high-level requirements for the kitchen.
2. Break these into subsystem requirements
3. Identify an integration challenge and propose a solution:

Now try the same task for a greenhouse.

:

Sources cited:

https://www.nasa.gov/reference/2-0-fundamentals-of-systems-engineering/
NASA Systems Engineering Handbook
NASA.gov
Enclothed cognition (2012) Hajo Adam, Adam D Galinsky

Disclaimer

This book is intended for educational and professional development purposes only. The content reflects the author's personal experience and opinions and shall not be construed as professional engineering services, design work, or technical advice for any specific project. Readers are encouraged to exercise their own judgment and, where applicable, consult a licensed professional engineer for technical matters requiring certification or design responsibility.

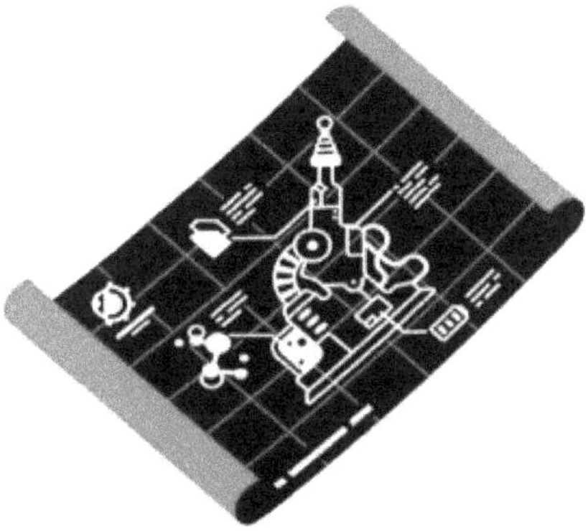

THE ELITE ENGINEER BLUEPRINT

www.ingramcontent.com/pod-product-compliance
Lightning Source LLC
LaVergne TN
LVHW051923051126
851857LV00021B/2117
9798234091383